Aakanksha Sadekar

Well planning report for the Molly field exploration well

GRIN Verlag

Imprint:

Copyright © 2011 GRIN Verlag GmbH
Druck und Bindung: Books on Demand GmbH, Norderstedt Germany
ISBN: 978-3-656-64194-0

This book at GRIN:

http://www.grin.com/en/e-book/271887/well-planning-report-for-the-molly-field-
exploration-well

School of Engineering
COURSEWORK SUBMISSION SHEET

All sections except the "LATE DATE" section must be completed and the declaration signed, for the submission to be accepted.

Any request for a coursework extension **must be submitted on the appropriate form (please refer to** http://www.rgu.ac.uk/academicaffairs/quality_assurance/page.cfm?pge=44250**), prior to the due date.**

Due Date	Date Submitted	For official use only
09/12/2011	09/12/2011	LATE DATE

MATRIC No. 1116462	
SURNAME- SADEKAR	
FIRST NAME(S)- AAKANKSHA	
COURSE & STAGE MSc Oil & Gas Engineering	Full Time
MODULE NUMBER & TITLE	ENM 201-WELLS
ASSIGNMENT TITLE	WELLS COURSEWORK
LECTURER ISSUING COURSEWORK	JOHN BISSET AND OWEN JENKINS

I confirm: (a) That the work undertaken for this assignment is entirely my own and that I have not made use of any unauthorised assistance.
(b) That the sources of all reference material have been properly acknowledged.
[NB: For information on Academic Misconduct, refer to
http://www.rgu.ac.uk/academicaffairs/assessment/page.cfm?pge=7088]

Sign- AAKANKSHA SADEKAR...Date.....09/12/2011..................

Marker's Comments

Marker	Grade

WELL PLANNING REPORT FOR THE MOLLY FIELD EXPLORATION WELL

LIST OF CONTENTS

1.0   Executive Summary:

The Kenmac Petroleum Corporation is planning to drill a new vertical exploratory well in their newly licensed Molly field. This report covers the detailed description of tasks that are required to be carried out for well planning.

☐ The Jack-up rig Trident II is selected. This selection is mainly based on the operability of the rig in the given conditions of the operating area i.e. water depth of the field, required derrick capacity, weather parameters etc.

☐ The casing scheme is based on the casing diameters those are minimum feasible given the formation evaluation requirements and drilling and production equipment sizes.

☐ The mud programme is selected after giving due consideration to the various formations to be drilled and the required hydrostatic pressure balance required to maintain the pore pressure of the reservoir.

☐ Cement design takes in to consideration the presence of "Zechstein" formation which is known mobile/plastic (salt) formation.

☐ Cased-hole Cemented Completion is proposed.

Overall the well design is to provide the maximum economic value of deliverability, serviceability and functionality for the target reservoir(s) to be penetrated. It meets the need for reliable pressure containment and well integrity over its life.

2.0 INTRODUCTION:

Molly is newly licensed field located in "Kenmac Petroleum Corporation's" North Sea Block 14/ab and is planning to drill vertical exploratory well within the "Carboniferous Siltstone" reservoir at 10000ft true vertical depth (TVD), below rotary table (BRT).

In this well drilling programme, following topics are covered in detail:

- selection of appropriate rig,
- estimation of pore and fracture pressure gradients,
- potential drilling problems,
- casing scheme,
- mud program,
- cementing programme,
- design and estimation of derrick capacity,
- well safety
- completion design.

The objectives of this well drilling programme are to:

I. Drill, and complete the well without any safety or environmental incidents
II. Drill each section with one bit shoe to shoe.
III. Achieve good quality cement jobs on each size of casing strings.
IV. Drill the reservoir section without any problems and obtain good quality data.
V. Perforate and complete the well in "Carboniferous siltstone" reservoir with minimal formation damage.

3.0 ASSUMPTIONS:

The following general assumptions are made to complete the well planning process. However, some more specific technical assumptions are also required in relevant sections which are stated therein separately:

1. Permits and approvals from the regulatory agencies have been obtained prior to rig arrival on the location.
2. Seabed survey and shallow gas studies are carried out.
3. 30" Conductor casing pre-set at 300m TVD BRT.
4. Average drilling rate over the entire well is 6.25m/hr (20.5ft/hr), including trips.
5. Studies to determine the pore and fracture pressures of the formations are carried out.
6. Corrosive gases such as H_2S and CO_2 are absent in the reservoir.

4.0   Basis of Well Planning and Design:

The following individual programs are included in the well planning and design of the well to be drilled in the field.

4.1 Drilling Rig Selection:

The drilling rig is selected to suit specific well drilling requirements. Following considerations are taken in to account to select the drilling rig:

- Formation pressures.
- Proposed casing scheme.
- Drill string size and weight(s) to be used.
- Required derrick capacity
- Hydraulic requirements.
- Any auxiliary equipment requirements.

From Transocean's website the suitable rig to drill this well is Jack-up rig "Trident-II". Detailed specifications are at **Appendix A**. Options of using semi-submersible and platform rigs were considered. Ultimately those options were discounted due to semi-submersible will be limited by water depth of the field and an expensive option. Since there are no concrete plans as yet for production from this well, installation of platform will not be there, hence, no platform rig.

4.2 Pore and Fracture Pressure Gradient Estimation:

Pore and fracture pressure estimations play an important role in successful planning and drilling of a well. The impact on well safety and economics can be catastrophic if these two important down hole pressures are not correctly estimated in well planning stage.

The low side prediction is used for consideration of losses and sticking. The expected prediction is used for mud weight determination. The high side prediction is used for selecting casing setting depth, leak-off / limit test requirements, kick tolerance and well control etc.

Pore and fracture pressures can be predicted from the following:

1. Log data or d-exponent data from an offset well.

2. Density logs from offset wells.

3. Using Eaton's equation.

Table 2 below shows the expected pore and fracture gradients in ppg and psi/ft:

TVD BRT (ft)	Pore Pressure (ppg)	Pore Pressure (psi/ft)	Fracture Pressure (ppg)	Fracture Pressure (psi/ft)
400	9	0.468	12.5	0.650
1500	9	0.468	13.5	0.702
4200	10.5	0.546	13.5	0.702
5000	10.5	0.546	14.5	0.754
7980	11.8	0.614	15.5	0.806

Table 2 –Molly Field Pressure Profile Plot

The pore and fracture pressure profiles are plotted in following graph acted nd the same plot is also used for estimation of casing setting depths by accounting for trip and kick margins of 0.5ppg respectively.

Molly Field - Pressure Profile & Casing Seat Selection

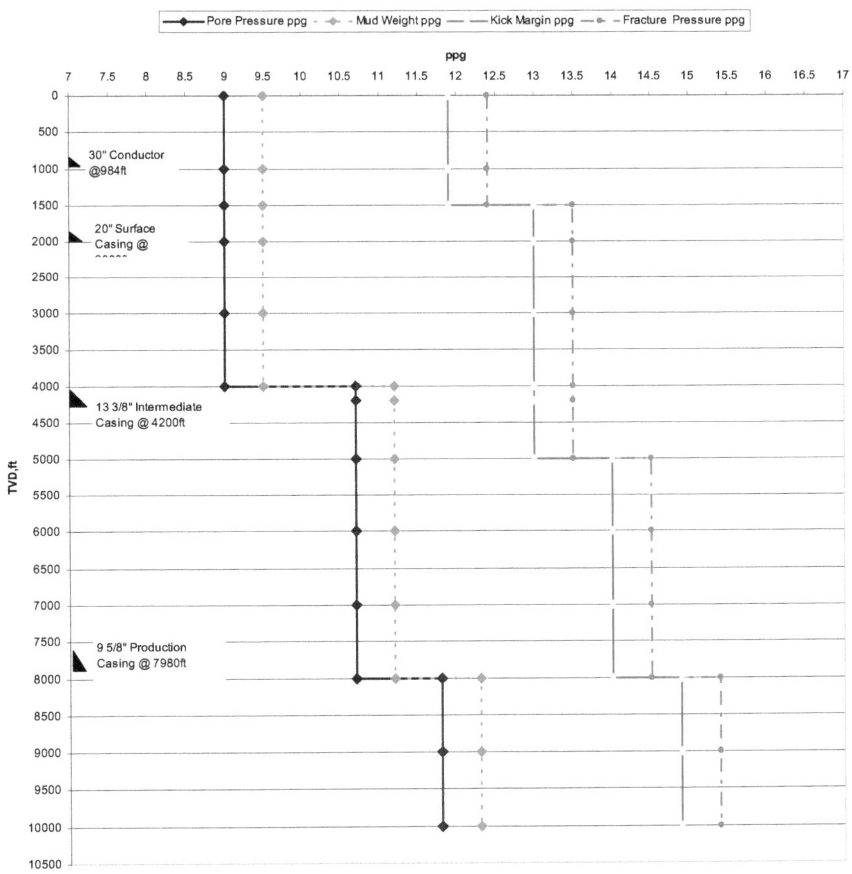

4.3 Anticipated Drilling Problems / Lithological Characteristics:

The Molly field lithological column suggests carboniferous sedimentary depositions with each of the layers may pose its own peculiar drilling problems.

Table 3 below lists some of the anticipated drilling problems which can encounter while drilling the well through these formations.

Geological Prognosis	Anticipated Drilling Problems	Mitigation
Sea-bed – 650 ft. Silts & soft clays, Unconsolidated clays	• Hole Problems – Unconsolidated clays form Gumbo at surface	• Drill section with seawater sweeps
650-4260 ft. Paleocene middle shales, clays and light sandstones.	• Formation water migrations may occur at the top ends of the Paleocene middle shales – Shallow water kicks can be expected • Wellbore instability due to reactive shales – stuck pipe	• Diverter will be installed from start of drilling. • Pre-treatment with suitable filter cake enhancing additive, combined with good drilling practices will be the main methods for avoiding stuck pipe
4260-5400 ft. Paleocene lower sands.	• Seepage losses	• LCM (Loss Circulation Material) treatment in mud
5400-5910 ft. Danian, chalks and fine / medium sandstones.	• Loss zone • Negligible window for mud weight	• LCM (Loss Circulation Material) treatment in mud
5910-7260 ft. Upper and lower Cretaceous, chalk /limestones.	• Loss zone • Negligible window for mud weight	• LCM (Loss Circulation Material) treatment in mud

7260-7980 ft. Upper Jurassics, shales / sandstones leading at lower end to fine / medium "Piper" sandstones.	• The Sandstones in this formation are known to be abrasive which can lead to BHA wear and possible low rate of penetration.	• Solids control in mud
7980-9765 ft. Triassic, shales and possible fluid-bearing sandstones on post Zechsteins.	• Zechstein salts are mobile - Stuck pipe, Casing collapse are main risks	• Salt saturated mud • SOBM • Stuck pipe prevention plan in place • Jar placement in BHA
9765-9960 ft. Permain Zechsteins, anhydrite and dolomites.	• Zechstein salts are mobile - Stuck pipe, Casing collapse are main risks	• Salt saturated mud • SOBM • Stuck pipe prevention plan in place • Jar placement in BHA
9960-TD Carboniferous siltstones/shales.	• Potential for kick is high	• Well control measures

Table 3 – Anticipated Drilling Problems

4.4 Casing Programme:

Casing pipes are lowered in a well with following objectives:

- to hold unconsolidated formations in place;

- to isolate high pressure zones from low pressure;

- allow production/injection of fluids to and from well.

With the above in mind the casing setting depths for Molly field are selected by comparing formation strength with anticipated pressure loads to which the formation may be subjected.

Table 4 below shows the proposed casing scheme:

Casing Type	Casing Size (OD) (Inches)	Hole Size / Bit size (Diameter) (Inches)	Casing Steel Grade	Casing Setting Depth TVD BRT (ft)
Conductor	30″	Pre-set	X-56	984
Surface Casing	20″	26″	J-55	2000
Intermediate Casing	13 3/8″	17 ½″	N-80	4260
Production Casing	9 5/8″	12 ¼″	L-80	7980
Liner	7″	8 ½″	L-80	10000/TD

Table 4 – Casing Scheme

30″ Conductor – Pre-set to cover the unconsolidated clays up to 650ft.

20″ Surface Casing – As per the Molly field pressure profile, the next ramp in pore pressure is seen from the depth at 4260ft. It will therefore need higher mud weight to drill the next hole section beyond 4260ft and hence there will be requirement to case off before 4260ft. Also the length of the hole section between 984ft to 4260ft becomes 3276ft long, which will take 159hrs to drill it with an average drilling rate of 6.25m/hr (20.5ft/hr). It is quite a long time (159hrs) to keep Paleocene middle shales open with water based mud and might have swelling and falling tendencies and hence need to be isolated. The 20″ surface casing is therefore proposed to be set at 2000ft.

The further changes in pore pressures and mud weight in the given pressure profile dictate the setting depths of the intermediate and production casing strings.

4.5 Mud Programme:

The primary function of the mud is to remove drilled cuttings from the well and to prevent formation fluids from entering into the well while drilling.

The two types of commonly used mud systems in the industry are:

- Water based

- Oil based.

Water based has water as continuous phase whereas oil is for oil based mud. Sometimes air is also used as a drilling fluid but its use is limited to areas where the formation is competent and impermeable. In mature or depleted reservoirs gas-liquid mixture (foam) is also used as drilling fluid.

Water based mud cause instability in drilling through shales due to hydration of clays present in shales. This phenomenon gets aggravated further in inclined or directional wells. In view of this water based muds are used in drilling of vertical or slant wells. Oil based mud do not contain free water and hence do not react with the clays in shale, thus making it excellent for use in formations where reactive shales are present. Oil based mud is also used to drill ERD (extended reach drilling) type wells

Table 5 shows the suggested mud programme:

Type of Formation (ft)	Hole Size / Casing Diameter (inches)	Range of mud weights for selection (ppg)	Selected Mud Weight (ppg)	Type of Mud
650 – 2000, Paleocene middle shales, clays and light sandstones	26"/20"	9.5 – 13.0	10.0	Water based mud, Seawater, Bentonite, KCl
2000 – 4260, Paleocene middle shales, clays and light sandstones	17 ½" / 13 3/8"	9.5 – 13.0	10.0	Water based mud, Fresh water, Bentonite, KCl
4260 – 7980, Paleocene lower sands, Danian, chalks and fine / medium	12 ¼" / 9 5/8"	11.0 – 14.0	11.5	Water based mud with KCl and PHPA additives

sandstones, upper and lower cretaceous chalk / limestones, Upper Jurassics, shales / sandstones leading at lower end to fine / medium "Piper" sandstones				
7980 – TD, Triassic, shales and possible fluid-bearing sandstones on post Zechsteins, Permian Zechsteins, anhydrite and Dolomites, Carboniferous siltstones / shales	8 ½" / 7"	12.3 – 15.0	12.5	Salt saturated Water based mud

Table 5 – Mud Programme and chosen mud weights to drill each section

The surface hole will be drilled with bentonite-seawater solution. This will help to drill unconsolidated clays by dissolving it in the solution and avoid gumbo formation at surface. The 12 ¼" hole section will be drilled by water based mud using KCl and polymer additives to control reactive shales from swelling and ensure trouble free drilling. The 8 ½" hole section will drill through Zechstein where bit and stabilizer sticking is common due to the salt "relaxing" into the well bore as it stress relieves. This risk will be mitigated by using water-based muds as there is an inevitable amount of washing-out that occurs. If squeeze does occur, a fresh water pill, viscosified and weighted to system density will be pumped across and spotted at the sticking zone. This pill will quickly dissolve the salt around the string, thereby freeing the pipe.

4.6 Cement Programme:

The objective of the casing cementation is to provide sealing to the casing annulus with outside formation and isolation of two different zones in a well. The cement bond also acts as a barrier to formation fluids from entering in to the well.

Table 6 below shows the proposed cementing programme:

Type of Casing, (Depth TVDBRT)	Type of Cement	Proposed additives	Remark
30" Conductor @ 984ft	-	-	• Pre-set
20" Surface Casing @ 2000ft	Class G Lead 12ppg Tail 16ppg	Foam Preventer, Extender, Accelerator	• Cemented to surface • 100% excess volume over the theoretical hole volume is estimated.
13 3/8" Intermediate Casing @ 4260 ft	Class G Lead 12ppg Tail 16ppg	Foam Preventer, Extender, Retarder, Dispersant, Bonding Agent	• Theoratical Top of cement planned at 3000ft • Extender will use in lead slurry to lower the cement density.
9 5/8" Production Casing @ 7980ft	Class G Lead 12ppg Tail 16ppg	Foam Preventer, Extender, Retarder, Dispersant, Bonding agent, Fluid loss agent KCl	• Theoratical Top of cement planned at 7000ft • Extender will use in lead slurry to lower the cement density. • KCl will be used in Tail slurry.

7" Production Liner @ 10000ft	Class G Lead 16ppg	Foam Preventer, Retarder, Dispersant, Bonding agent, Fluid loss agent	• Theoratical Top of cement planned at Top of Liner @ 7780ft

Table 6 – Cementing Programme

The planned top of cements for 13 3/8" casing and 9 5/8" casing are not up to surface for following reasons:

- Total volume of cement slurry to be pumped becomes in both cases will be too large to pump in one stage. Thus entire cement job may exceed the designed thickening time of cement slurry. It can thus pose a risk of premature setting of cement slurry while pumping.
- To deal with thermally induced annulus pressures behind the casing string. Vacant annular space provides vent to these pressures.
- Keep a room to cut and pull casing in future for any potential sidetrack of the well.

4.7 Derrick Capacity:

The derrick capacity calculations are shown in **Appendix D**. Based on those calculations the required derrick capacity is 847084lb. The rig Trident-II has derrick capacity of 1,245,825lbs (565mTons) and hence is suitable rig for Molly field operations.

4.8 Well Safety:

Safety of the personnel on the rig, equipment and environment are top priorities while drilling a well. In order to maintain well safety, following procedures and equipment are used:

- Allowances for trip margin and kick tolerance are applied over pore and fracture pressures to choose mud weight.
- The Blowout Preventer (BOP) equipment rating is designed and selected to meet the highest anticipated pressures during drilling of the well.
- Sub-surface safety valve (SSSV) and annulus safety valve (ASV) are installed in completion tubing.

- Casing design is carried out to meet the anticipated pressure loads and mud programme is designed to balance them.
- Cement jobs are designed and carried out to place enough cement in annulus to cover the reservoir formation.

The BOP is an important piece of equipment used in well control. The main function of the BOP is to shut the well bore and stop it flowing in case of loss of primary control, and to be able to maintain bottom hole pressure equal to the formation pressure while gaining back the primary control.

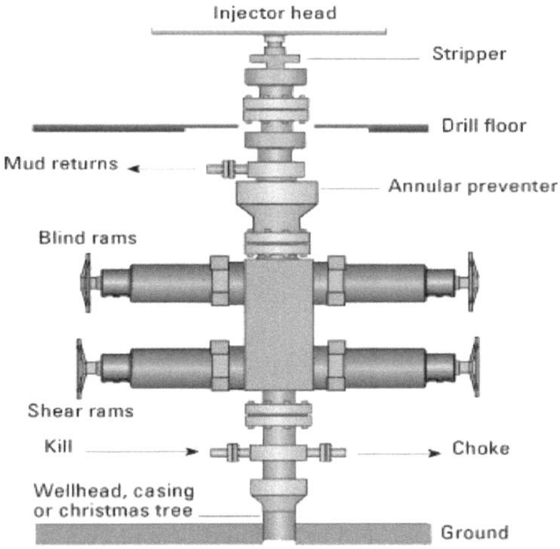

Fig. - Blowout preventer (Schlumberger Limited 2010)

The standard pressure rating for BOP has been worked out as 10,000 psi as shown in **Appendix E**. The BOP on Trident-II has 10000 psi pressure rating and meets the requirement.

4.9 Completion Design:

The various down-hole equipment or procedure required to be carried out for bringing the well fluid to surface is called well completion.

Wells are either completed by single or dual completion methods:

Dual completion:

It is used to produce from two or more formations simultaneously. The dual completion comprises of minimum two packers and tubing strings.

Single completion:

Single completions are used to produce from single zone only. This is cost effective than dual completions.

Since Molly field production is proposed from single reservoir only hence single completion is proposed. As a minimum following down-hole equipment and or procedure will be used:

Down-hole Equipment / Procedure	Purpose
Perforations	• To establish fluid communication between reservoir and well bore.
Subsurface Safety Valve	• To shut the well in controlled manner.
Production Tubing	• It is used as a medium to bring well bore fluid to the surface.
Packers	• Acts as a seal in between tubing and casing or liner.
Sliding Sleeve / Circulating Sleeve	• It gives a passage to tubing and annulus for circulation/injection • Used for selective production from two different zones.
Landing Nipple	• To provide profile for installation of flow control device

4.10 Well Schematic:

Proposed well schematic for Molly field well is as given below. It details the casing, cementing and completion details.

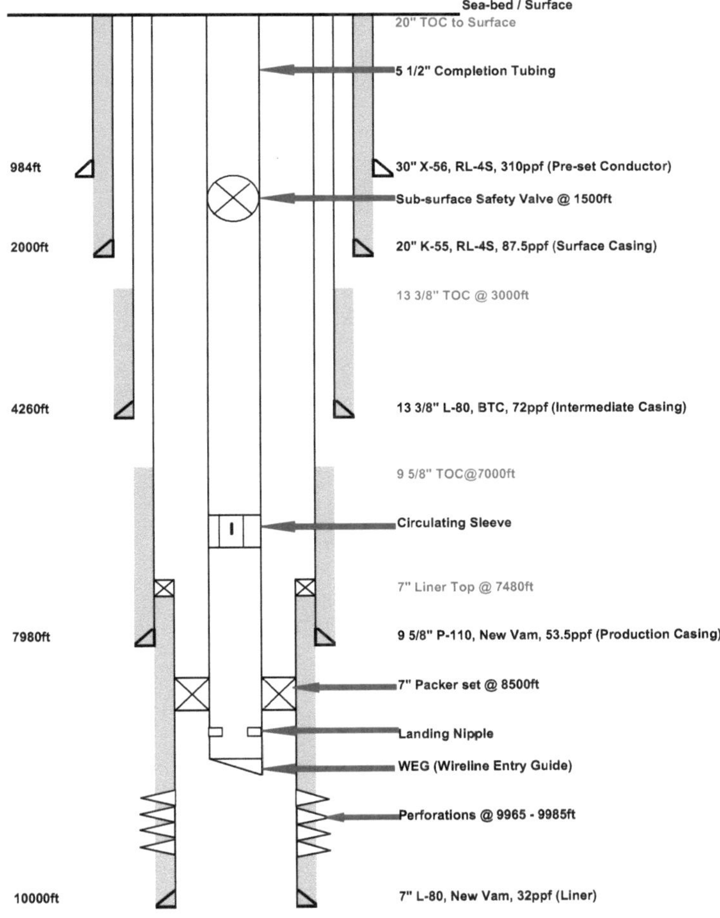

5.0   Conclusion:

The well design of the Molly field exploration well has taken in to account all aspects of well life cycle. The proposed design is cost effective and each section of the well is designed to meet this conservative approach. While the well design is based on the aim of minimising the cost the well safety is not compromised at any stage.

6.0   References:

Bourgoyne, A. et al., 1986. *Applied Drilling Engineering, Vol.2.* Richardson, TX: Society of Petroleum Engineers.

Economides, M., Watters, L., Dunn-Norman, S. 1998. *Petroleum Well Construction.* Chichester: Wiley.

Transocean Ltd, 2011. *Fleet Overview,* available from: http://www.deepwater.com/fw/main/Trident-II-31C16.html?LayoutID=17 [Accessed 5 December 2011]

Gatlin, C. 1991. *Petroleum Engineering: drilling and well completions.* Englewood Cliffs, NJ: Prentice Hall.

Atul S. Bhadauria – Personal communication 4th December 2011

Avinash Sadekar – Personal Communication 30th November and 6th December 2011.

Book on Drilling Engineering: Department of Petroleum Engineering, Heriot-Watt University

INSTITUT FRANÇAIS DU PÉTROLE PUBLICATIONS., 1999. Drilling data handbook. 7th ed. Paris: Éditions Technip.

7.0   Appendices

Appendix A: Specifications of Jack-up Rig "Trident-II"

Trident II

The TRIDENT II is a 3 leg Marathon Le Tourneau 116C cantilever Jackup capable of operating in water depths up to 300 feet. A 13 5/8 in 10,000 psi BOP is used to obtain maximum drilling depths up to 25,000 feet.

Rig Type - Non-U.S. Jackups

Design - MODU (JDU) Marathon LeTourneau 116C

Builder - Marathon LeTourneau Singapore. Refitted

Year Built - 1977/1985

Classification - ABS Class A1 Self Elevating Drilling Unit

Flag - Panama

Accommodation - 89 berths

Helideck - Rated for Sikorsky S-61

Max Drill Depth - 25,000 ft / 7,620 m

Max Water Depth - 300 ft / 91 m

Operating Conditions - Wind: 50 knots; Waves: 28 ft at 12 seconds; Current: 0 knots

Storm Conditions - Wind: 100 knots; Waves: 44 ft at 15 seconds; Current: 1 knots

Technical Dimensions

Length - 237 ft/72 m

Breadth - 200 ft/61 m

Depth - 26 ft/8 m

Ocean Transit Draft - 27 ft/8 m

VDL - Operating 1,609 st/1,460 mt

Capacities

Liquid Mud - 2,300 bbls/12,913 cu ft/365 cu m

Drill Water - 14,918 bbls/83,757 cu ft/2,370 cu m

Potable Water - 1,725 bbls/9,685 cu ft/274 cu m

Fuel Oil - 2,718 bbls/15,260 cu ft/432 cu m

Bulk Mud - 4,000 cu ft/113 cu m

Bulk Cement - 4,000 cu ft/113 cu m

Sack Material - 5,000 sacks

Drilling Equipment

Derrick - Lee C. Moore 160 ft x 30 ft x 30 ft, 565 mt nominal capacity

Drawworks - Gardner Denver 2100 E, input horsepower: 3000

Top Drive - Varco TDS-4S

Rotary - National Oilwell, 37½ in opening

Mud Pumps - 2 x Gardner Denver PZ-11 (1600hp)

Shale Shakers - 1 x Brandt scalping; 3 x Derrick Flo Line

Desander - 1 x Crestex w/ (3) 12 in cones (1500 gpm)

Desilter

Mud Cleaner - 1 x Derrick Mdl 48 mud cleaner w/ (20) 4 in cones

BOP - 1 x Hydril 13-5/8 in 5,000 psi annular, H2S Trim; 1 x Hydril 21-1/4 in 2,000 psi annular, H2S Trim; 2 x CIW (single, double) 13-5/8 in, 10,000 psi

Diverter - Hydril MSP 29½ in, Pressure rating: 500 psi

Control System - Koomey hydraulic control system

Choke & Kill - CIW 4-1/16 in 10,000 psi, H2S Trim

Cementing - Dowel Schlumberger SCS-R300; 10,000 psi unit

Machinery

Main Power - 5 x CAT D399 TA (1195 hp); Generator: GE 5AT AC, 600 V, 60 Hz, 1200 kVA, 930 kW

Emergency Power - 1 x CAT D398 (1000 hp); Generator: GE AC, 600 V, 60 Hz, 750 kVA, 500 kW

Power Distribution - SCR system:GE; Rated at: 1200 amps and 750 volts

Deck Cranes - 3 x ML PCM-120-AS; 45 mt at 24 ft, 9 mt at 85 ft maximum radius

Jackup Specifications

Legs - 3 x 410 ft

Spud Cans - Each leg equipped with 1585 sq.ft. spud can

Jacking System - Marathon LeTourneau rack & pinion

Cantilever/Slot Cantilever

Mooring Equipment

Winches - 4 x LeTourneau W 1500 Wire/Chain

Anchors - 4 x 4.5 mt anchors

Appendix B: Computation of Mud Weight ranges

For calculation of mud weight ranges a trip margin and kick tolerance of 0.5ppg is considered over pore and fracture pressures.

1. Sea-bed-4260ft
   Pore pressure = 9.0ppg
   Fracture Pressure = 13.5ppg
   Pore pressure with trip margin = 9.0+0.5 = 9.5ppg
   Fracture pressure with kick tolerance = 13.5-0.5 = 13.0ppg
   Proposed range of mud weights = 9.5 – 13.0ppg

2. 4260ft-5000ft
   Pore pressure = 10.5ppg
   Fracture Pressure = 13.5ppg
   Pore pressure with trip margin = 10.5+0.5 = 11.0ppg
   Fracture pressure with kick tolerance = 13.5-0.5 = 13.0ppg
   Proposed range of mud weights = 11.0 – 13.0ppg

3. 5000ft-8000ft
   Pore pressure = 10.5ppg
   Fracture Pressure = 14.5ppg
   Pore pressure with trip margin = 10.5+0.5 = 11.0ppg
   Fracture pressure with kick tolerance = 14.5-0.5 = 14.0ppg
   Proposed range of mud weights = 11.0 – 14.0ppg

4. 8000ft-10000ft (TD)
   Pore pressure = 11.8ppg
   Fracture Pressure = 15.5ppg
   Pore pressure with trip margin = 11.8+0.5 = 12.3ppg
   Fracture pressure with kick tolerance = 15.5-0.5 = 15.0ppg
   Proposed range of mud weights = 12.3 – 15.0ppg

Appendix C: Computation of Casing Grades and Weights:

The safety factor of 1.25 is considered for estimation of collapse pressure and 1.3 for burst pressure of casing strings.

1. 20" Surface Casing

   Hydrostatic pressure = MW*0.052*TVD

   = 10.0*0.052*2000

   =1040 psi

   Collapse pressure for casing = HP*1.25

   = 1040*1.25

   = 1300 psi

   Burst pressure for casing = HP*1.30

   = 1040*1.3

   = 1352 psi

2. Intermediate Casing (13 3/8") :

   Hydrostatic pressure (HP) = MW*0.052*TVD

   = 10*0.052*4200

   =2184 psi

   Collapse pressure for casing = HP*1.25

   = 2730 psi

   Burst pressure for casing = HP*1.30

   = 2839 psi

3. Intermediate Casing (9 5/8") :

   Hydrostatic pressure (HP) = MW*0.052*TVD

   = 11.5*0.052*7980

   =3285 psi

   Collapse pressure for casing = HP*1.25

   = 5965 psi

   Burst pressure for casing = HP*1.30

   = 6204 psi

4. Production Liner (7") :

   Hydrostatic pressure (HP) = MW*0.052*TVD

   =12.5*0.052*10000

$$=6500 \text{ psi}$$

Collapse pressure for casing = HP*1.25

$$=8125 \text{ psi}$$

Burst pressure for casing = HP*1.30

$$= 8450 \text{ psi}$$

Based on the collapse and burst pressures as calculated above, the grade and nominal weight of casings, including threads and couplings are selected from API tables as following:

1. Surface Casing:

   Grade- K-55

   Nominal Weight – 133lb/ft

   Total weight of surface casing = 133*2000

   $$= 26600 \text{ lb}$$

2. Intermediate Casing:

   Grade- L-80

   Nominal Weight – 72lb/ft

   Total weight of intermediate casing = 72*4260

   $$= 306720 \text{ lb}$$

3. Intermediate Casing:

   Grade- P-110

   Nominal Weight – 43.50 lb/ft

   Total weight of intermediate casing = 43.5*7980

   $$= 375060 \text{ lb}$$

4. Production Liner:

   Grade-P-110

   Nominal Weight – 32lb/ft

   Total weight of Production Liner = 32*2500

   $$= 80000 \text{ lb}$$

Appendix D: Estimation of Maximum Hook Load applied to Derrick Structure:

Casing Weight - The estimation of the maximum hook load applied to derrick or derrick capacity is determined by taking the maximum weight in between casing and drill string for the well.

As calculated in earlier section the maximum casing weight to be handled during drilling of the well is the weight of the 9 5/8" casing. It is 375060lb.

Weight of Drill String - 5", G-105, 19.5 ppf, drill pipe is to be used, as the rig Trident-II has this available. Apart from the 5" DP, the drill string is also comprised of 6 ½" drill collars and 5" heavy weight drill pipe (HWDP).

The number of drill collars required is calculated from,

$$\text{Dc Length} = \frac{1.15 \times \text{WOB}}{\text{BF} \times \text{Wdc}}$$

Where,

15% safety factor = 1.15

WOB = Weight on bit. Assumed as 5000lbsf/inch of bit OD

BF = Buoyancy Factor

Wdc: Drill Collar weight in air

$$\text{Dc Length} = \frac{1.15 \times (5000 \times 10.625)}{\{1 - (10/65.5)\} \times 91.59}$$

So, Dc Length = 787.22ft,

The length of drill collar is normally 31ft so the total number of drill collars required will be = 787.22/31 = 25.39 ~ 26 drill collars

It is assumed that HWDP will be half of the total drill collars required.

So Number of HWDP required = 26/2 = 13

Let us say the total weight at surface coming on the top joint of drill pipe is denoted by Tsurf,

Then Tsurf = [(Ldp x Wdp) + (Ldc x Wdc) + (Lhwdp x Whwdp)] x BF

where,

Ldp: Length of drill pipe,

Wdp: Weight of drill pipe per unit length,

Ldc: Length of drill collar,

Wdc: Weight of drill collar per unit length,

Lhwdp: Length of HWDP,

Whwdp: Weight of HWDP per unit length,

BF : Buoyancy Factor

So Tsurf = [{10000 – (31 x 26) – (31 x 13)} x 19.5 + {(31 x 26) x 91.59}
+ {(31 x 13) x 57.93}] x {1 – (10/65.5)}
So Tsurf = 88497.21lbs ~88500lbs.

From above calculations it can be seen that the weight of the casing is going to be the maximum weight which rig's hoisting system will have to handle and hence it is used to calculate the capacity of the derrick.

It is assumed that crown block sheave has 10 lines and it will be used to calculate the derrick capacity. n = 10
Assuming over-pull capability = 150000lb and top drive weight = 80000lbs with 1.1 as safety factor, the total weight W will be:
= (375060 + 150000 + 80000) * 1.1
= 605060lb

The derrick capacity is calculated by using following formula:

$$F_d = \left\{ \frac{1 + E + En}{En} \right\} * W$$

Where, W= Total Weight = 605060lbs, n=10, E=0.810

So, F_d = 740265lbs

The equivalent load is calculated by following formula,

$$F_{de} = \left\{ \frac{n+4}{n} \right\} * W$$

Thus, $F_{de} = 1.4 * 605060 = 847084 \text{lbs}$

Hence, the design derrick capacity will be 847084lbs.

Appendix E: Estimation of BOP standard Pressure Rating

Assuming reservoir gas gradient = 0.1psi/ft.

Thus, the gas kick can exert a pressure, P_g = 0.1*10000 = 1000psi

Let the maximum fracture pressure in Molly field be denoted by P_h

= P_h

= 15.5*.052*10000

= 8060 psi

where 15.5ppg is the maximum fracture pressure between 8000ft-10000ft.

So Standard Pressure Rating for BOP = P_h – P_g

= (8060 – 1000) psi

= 7060 psi

Thus, the standard BOP pressure rating for the BOP will be 10000psi.